BLUFF

THE QUANTUM UNIVERSE

JACK KLAFF

RAVETTE PUBLISHING

Published by Ravette Publishing Limited
P.O. Box 296
Horsham
West Sussex RH13 8FH

Tel: (01403) 711443
Fax: (01404) 711554

© Oval Projects 1997
All rights reserved, including the right
of reproduction in whole or in part
in any form.

Series Editor – Anne Tauté

Cover design – Jim Wire, Quantum
Printing & binding – Cox & Wyman Ltd.
Production – Oval Projects Ltd.

The Bluffer's Guides™ series is based
on an original idea by Peter Wolfe.

The Bluffer's Guides™, Bluffer's™
and Bluff Your Way™ are Trademarks.

An Oval Project
for Ravette Publishing.

Dedicated to:
Anna, Nina and Kaz.

CONTENTS

Preparing Yourself   7
    The Big and the Small   7
    The Paradox of the Belief System   8
    Two Schools of Thought   9
    The Language Barrier   11
    Tactics for Tight Corners   13

The Quantum Realm   15
    The Quantum Moment   15
    The Medium and The Message   16
    Quantum Mechanics   18
    Matter   19
    Quarks and The Standard Model   22
    Antimatter   24
    The Quantum Leap   25

The Shocks   27
    Causality   28
    Predictability   29
    Reversibility   30
    Continuity   31
    Accurate and Informative Measurement   33
    Objectivity   35
    Locality   37
    Order   39
    Clear Definition   40
    Separateness   41
    Either/Or Thinking   43
    Certainty   43

Cooking Your Own GUT   44
    Stepping Towards a TOE   45
    Superstrings and M-Theory   47

Famous Physicists 49
- David Bohm — 49
- Niels Bohr — 49
- Max Born — 50
- Louis-Victor de Broglie — 51
- Paul Adrien Maurice Dirac — 51
- Paul Ehrenfest — 51
- Albert Einstein — 52
- Enrico Fermi — 52
- Richard Feynman — 53
- Murray Gell-Mann — 53
- Stephen Hawking — 54
- Werner Heisenberg — 54
- Max Planck — 55
- Wolfgang Pauli — 55
- Ernest Rutherford — 56
- Erwin Schrödinger — 56

The Implications 57
- What it All Means — 57
- The Cutting Edge — 58
- To Be Continued… — 60

Glossary 61

INTRODUCTION

Whether you are trying to explore the universe, which is very, very big, or the quantum realm in which particles of light and matter are very, very small, nothing can be stated with clearer conviction than this: nobody understands what's going on.

The most honest physicists – and, it has to be said, all the most brilliant ones – are in complete agreement; nature and the universe are astonishing, can appear absurd, and are known to flummox the best of them.

So, if you have lamented your own incomprehension of those entities which are very, very small and that which is very, very big, let yourself off the hook immediately. Welcome to a bluffer's paradise, where no-one need feel afraid, nor should anyone feel stupid.

Of course, where the physicists do have knowledge and understanding, prudence dictates that astute bluffers gain some appreciation of it. Happily, such an appreciation may be obtained without formulæ, equations or fractions.

It may seem to be taking the word 'appreciation' too far, but you are going to have to feign, find or affirm some sense of wonder. This will not be too difficult. The universe is after all very large, and the smallest entities are extremely small and their conduct is unusual to say the least. So it's a good idea to exhibit awe. The approach adopted by the great Danish physicist, Niels Bohr, is a compelling one: he used to say that if people didn't find what he was telling them absolutely amazing, they weren't taking it in. Indeed, the subject is so amazing that such geniuses as Einstein, Bohr, de Broglie, Feynman, Fermi, Planck, Pauli, Heisenberg, Schrödinger, Schwinger, Witten and yourself haven't quite cracked it.

Nothing conveys the impression of humungous intellect so much as even the sketchiest knowledge of quantum physics, and since the sketchiest knowledge is all anyone will ever have, never be shy of holding forth with bags of authority about subatomic particles and the quantum realm without having done any science whatsoever.

After all, whole careers have been built on such deception, and, besides, what else is the activity of bluffing about?

PREPARING YOURSELF
The Big and The Small

Whichever way you approach discussions about the universe and about the quantum realm, the notion of scale will somehow always be present. But remember that it should be beneath you to be impressed by size: it is for other people to reel and gape. This is something with which you must appear to be familiar.

Physicists have naturally what bluffers must somehow acquire; the ability to drop phenomenally massive numbers into the conversation with an ease that is at once reassuring and awe-inspiring.

Where the tendency is towards 'big', you need to appreciate that a light year measures the distance you could cover in a year zooming at 186,283 miles per second, and that to visit the nearest galaxy to earth would take 2 million years at the speed of light. 12,000 million billion miles is pretty far, even for a commercial traveller. (And if you do go, don't drink the water.)

At the other extreme, you need to have some idea of what is meant by 'small'. All matter can be broken down into **atoms**. Atoms are small. They are smaller than affordable apartments in Manhattan, they are smaller than portions at the Ritz, they are even smaller than the chance that a politician will be honest. The full stop at the end of this sentence will be a tiny blob of ink about a millimetre wide which will contain close to four billion atoms.

Having taken that into account, a reminder is necessary. To a human being, or to a full stop, an atom is small. But to a **subatomic particle**, any one of those four billion atoms in that full stop is enormous. It would feel much like a speck of dust would feel if it were to look up at a building twice as high as the Taj Mahal.

Pointing to a full grain of coffee, or parmesan cheese, or beach sand, or, best of all, a millimetre of human skin may be a game well worth the candle.

The Paradox of the Belief System

Few forms of knowledge are so closely interwoven with human belief systems as are studies concerning the Quantum Universe.

When discussing the very small things – tiny particles of light and matter, or wrestling with the very big universe, your belief system, and the belief systems of the people whom you will be impressing, will be powerfully challenged.

Because you are dealing with fundamental entities in nature, it will be impossible to avoid disturbing the most basic beliefs of those around the table. The mere mention of a particle will raise the issue of consciousness in snails and how this relates to God.

Everything that anyone has ever held dear, every blind assumption, every hard-won prejudice may be horribly threatened. Clearly, this sort of thing should not be allowed; and, certainly, no-one else should have to witness it.

To prepare yourself for deipnosophy (the noble art of excelling at dinner parties), you need to know where you stand. Behind closed doors you will have resolved all those profound questions which, however slightly, might have upset your own world view. You will have absorbed this knowledge, and been shaken and stirred, but you will have managed to keep your cherished notions intact. Before you re-emerge to mingle with your friends, you will have discovered how to use the very same knowledge to provide you with an opposite, more pleasing interpretation.

It will be no effort at all to find some subatomic occurrence which has a bearing on religion, or politics, or marriage or morals, or art or azaleas, or music or marmalade, or snooker or sex. Each precious phenomenon must be seized upon, every analogy squeezed until it squeaks.

Two Schools of Thought

Even among physicists prejudices come into play. There are two distinct schools of thought; here's how you can see them coming.

1. The Holists

Members of this School are, broadly speaking, holistic. Holists are interested in the whole picture, except when it comes to the letter 'w'. As time goes on, more and more holists are likely to come out of the (w)ood(w)ork.

2. The Reductionists

Those who belong to this school are interested in the parts of the picture. All they want to know is 'How small?', 'How many?', 'How wide?', etc., as if the Quantum Universe was a watch which could be taken apart to the smallest screw to find how it all works.

Holists are generally interested in Quality. Reductionists are interested in Quantity.

Holists can see the universe and eternity in an atom. They have likened the universe to an enormous ball of string. No-one can see its end or its beginning, yet if you draw on the string at any part of the ball, the ball becomes tighter throughout. Try to pull one strand of it, and all of it changes. Holists say this involves a deeper reality than mere inter-relationship. For them, wherever you tweak the string, there is the entire universe because all of it changes. Everything is part of everything, every part is part of every part. There is oneness and only oneness.

Reductionists concentrate on the individual atom without relating it to anything else. They think you should chop up the string ball and measure the bits.

Holists think that the scientific discoveries of the 20th century should encourage us to rethink everything. This is sometimes called a 'paradigm shift' – the altering of a model, in this case the whole model. They want to unite, to bind, to join. Reductionists don't. They like the word 'discrete', meaning 'separate from everything else'.

Reductionists think it is unnecessary to make so much of the fact that the iron at the core of planets is also contained in human hæmoglobin. Holists think it's poetic and meaningful that human beings have stardust in their blood.

Holists find coincidences quite intensely moving; reductionists demonstrate why they aren't coincidental. Holists accept a certain floppiness in their ideas; reductionists insist on rigour. Holists quite like the circuitous approach. A reductionist will even do a cutting gesture while explaining that space itself is curved so you will get that straight.

The Language Barrier

When you come up against the limits of the Quantum Universe, and indeed the limits of your own understanding, you will inevitably come up against the extremes of what you can express linguistically, so to speak.

For instance, the notion that in the beginning everything was in one place, and, lo, that place was nothing. No thing. It was a nothingness so nothing-like that its nothingness took nothingness to untold extremes of nothingness. (Modern physics, like love, or football, or being a teenager, defies language.)

It has never been easy for anyone to find words to explain the deep mysteries of the universe. How to explain new phenomena? How to describe them? How to tell people about entities and occurrences which are so far removed from anything hitherto known about or understood that the brain starts doing backflips and wants to join the circus.

The difficulties begin with names. You will, as it happens, encounter many words in the quantum realm which end in '-on'. There are hadrons – which may be baryons or mesons – and mesons which may be pions or kaons, while muons and tauons are leptons, not to be confused with sleptons, and so on.

You may be tempted to sort them out. Don't. At the University of California at Berkeley, in an almanac which records new particles, the list already runs to more than 2,000 pages. **Enrico Fermi** said if he could remember all the names given to particles he would have been a botanist, and if such things really do interest you, you need to get out more.

Niels Bohr, who used to have days of intense discussion about almost every word he used, was called a bad lecturer, but often he just wanted to get the language right. He once paused for a long moment before saying," ...and", then another very long pause before he said, "...but". (It was he who gave a French ambassador the happy and fulsome greeting: "Aujourd'hui!") At least he conceded that there was a problem. It took him several weeks of hard thought and discussion to come up with this: "We are somehow suspended in language".

You could do no better yourself.

Blaming language is your parachute, your cavalry, your knight. A reminder about linguistic difficulties will send sharp questions ricocheting back to their posers; and when you have dropped yourself in it, talking about the lack of suitable words can soften the horror-movie eyes around the table.

The language used by physicists is mathematics – even though many of them need help with it from real mathematicians. Their best-known trick is to shut their eyes and find some Greek letters to make it seem deep and meaningful. Physics is chock-full of words which have been poached and which may once have had a sporting chance of meaning something. Most could not be more confusing if they were designed to be so, and the suspicion is that they were.

It may also be a useful gambit to cite languages other than your own, such as Algonkian spoken by the Blackfoot people of Alberta, Canada, which emphasises change and movement in nature, rather than finding names for things. For example, instead of saying "I saw someone long ago", they say "I saw him/her in the far away". This so impressed the eminent scientist David Bohm, he proposed an entirely new kind of language should be developed for the quantum universe. An English one, of course.

Tactics for Tight Corners

Whenever the universe becomes a topic of conversation certain questions will always be hovering around the edges – if there are any edges, and that's one of the questions. These include:

- How did the universe begin?
- Who cares?
- If the universe had a beginning, will it have an end? (Or a new series next autumn?)
- What has 'quantum' got to do with it?
- I sort of get it, but just when I do, I lose it again. It's as if it's in the cracks of my understanding. Is that where it should be?
- How about a Bloody Mary?

Some of these penetrating queries may seem to be unanswerable. And that is because they are. But defeatism is alien to science as well as to bluffing. And since the scientists who make so bold as to deal with these questions are so often conjecturing, you may as well follow suit:

"Physicists now believe that before the universe began, everything was just bubble gum", or "Last week cosmologists declared that the universe was stacked between two bookends, now they're saying it's shaped like the upside-down trunk of a man in boxer shorts". Or, "The smallest particle, the carryon, is even smaller than the one found in February, the dreamon, and last Tuesday's equally minuscule particle, the whatson".

But remember to balance conviction with caution. Never commit yourself about the outer limits of the universe or the quantum realm even to a 'probably'. Anything you utter with certainty, or declare to be 'probably true' could return to haunt you and, it can be said with confidence, probably will. If you know what's good for you, a 'possibly' is the farthest you will go.

To avoid being thought a delinquent in the light of tomorrow's new data, you need to make the lack of language work for you by:

a) saying that we are up against the limits of it;

b) using the word 'possibly' as an insurance policy;

c) capping all statements about problems that have not been solved and theories which have been proved unassailable, or the finding of the smallest particle in the universe, with the caveat "So far".

With some of the more difficult questions, a gentle attack may be a good spirited defence. The Dismissal is ideal since it conveys the impression of very deep and thorough knowledge and prevents any deviation from your chosen conversational course. The Dismissal is the simple but salutary sentence: "The question doesn't apply."

This is particularly effective as part of a longer riff, as in: "The question doesn't apply, the question can't apply because we have to avoid ideas of 'there', indeed notions of 'because', because 'because' needs a 'dum de dum' in order to have gone from 'dum-de-dum' to 'dum-de-dee' because of 'blah-de-blah'. Which doesn't happen 'there' nor for that matter anywhere…"

But in most gatherings no answer conveys authority and wisdom like the sentence on its own, unadorned and cool as a padlock:

"The question doesn't apply."

THE QUANTUM REALM

The study of subatomic particles is called **quantum mechanics**. This is strange because the word 'quantum' is Latin for 'how much' or 'how great'. In this case, not a lot. And it can be seen within a nanosecond that it is entirely alien to this subject to talk of anything 'mechanical', 'mechanistic' or 'machine-like'.

The term 'quantum' was coined in 1900 by the great physicist, **Max Planck**. (Bluffers should avoid jokes about the name 'Planck'. The same advice applies to Bohr. And Schwinger.) The unit, the quantum which Planck devised, was to represent a 'package of energy'.

The Quantum Moment

Max Planck's professor advised him not to do physics because it was thought at that time that all its problems had been solved. Planck ignored his advice. In 1900 he found a problem to solve, and in solving it, set a million more in motion.

The problem was called, rather unfairly perhaps, the Ultra-Violet Catastrophe. It occurred as a result of 'black body radiation'. A black body – a black object – tends to absorb radiation when cool; but if you heat it sufficiently it will emit radiation by going red hot. The hotter it gets, the farther up the spectrum the glow issuing from it will get, through an orange and a blue heat, all the way to ultra-violet (the sort of light that picks out teeth and specks on a jacket in an old-fashioned discotheque).

For a physicist, seeing something work is not the point; the point is that the mathematics of it must work. The sums implied that the progress through the spectrum of the differently-coloured glows would be shattered in a burst of catastrophic ultra-violet radiation. But however intense the heat, no such destructive explosion occurred. Something was wrong with the calculations. So Planck did the obvious thing. He cheated. He squeezed a number into the sums, *viz*: 0.000 000 000 000 000 000 000 000 00066.

Then he attempted to do that most difficult thing for a scientist, he went back to real life. But whatever he tried, he could not get rid of that embarrassing little cheat.

It was some time before he realised that the cheat was the whole point. The cheat was Nature's cheat. The black body did not heat, or cool, in one smooth run of energy but in little jerks or spurts. (On reflection, perhaps Quantum is a better name than Jerk or Spurt – but only marginally.)

For years Planck's discovery was not taken by other physicists to be a revolution. Then when it was, it was precisely the kind of revolution he fought for the rest of his life to prove it wasn't. He had created a monster which threatened the very basis of classical science. Again and again he returned to the problem he had already solved, even making ready to undo his own work and diminish his ranking as a physicist. To remark that there never was a more reluctant springboard than Planck is a truly unintentional breach of the no pun rule.

The Medium and The Message

Quantum physics started off being concerned with the composition of matter, the properties of light and the interaction between light and matter. In 1900, when Planck was playing with fire and tinkering with his maths, what he was examining was the interaction between light and matter. The young **Einstein** explained in 1905 how photons – particles of light – could be directed at a metal and dislodge electrons – particles of matter. Bonus marks will come your way when you declare that Einstein's examinations of the photoelectric effect were indispensable when television was invented.*

Light is just one of a number of agents of transmission which dislodge, illuminate, attract, repel, bind or explode all the materials of nature in a constant dance. You will receive much thanks for the simple information that there are, basically, two main categories whereby subatomic particles may be classified:

1. particles of matter
2. messenger particles

The suspicion is that for each kind of matter particle there exists a 'symmetrical' messenger particle. If so, it would create a marvellous balance of medium and message throughout the universe which you should call **supersymmetry**.

It will probably do you no lasting harm to know that particles of matter have the generic name **fermions**, and all messenger particles, whatever else they are called, are also **bosons**. Saying that fermions are scared of commitment, individualistic and selfish, and that bosons are the outgoing, surrendering relationship particles, can be a delectable conversational gambit in mixed company.

---

*You can top this with the fact that Einstein won his Nobel Prize not for the General Theory of Relativity, nor for his big hair or his habit of not wearing socks, but for his work on the photoelectric effect.

Quantum Mechanics

You can divide the early history of the development of quantum mechanics into three main periods. During the first, the discoveries of Planck and Einstein were demonstrated but not fully understood. During the second, Niels Bohr's theory of 1913 was explored: his work with atoms and light was enormously helpful without, strictly speaking, being comprehensible. It was incomplete, which is why it was not until the third period that true quantum mechanics appeared when **Max Born**, **Werner Heisenberg**, **Louis V. de Broglie**, **Erwin Schrödinger**, **Pascual Jordan** and **P.A.M. Dirac** (among others) offered fuller descriptions of a body of physical law which not a single person on earth could understand.

There was a clear need for some outstandingly impressive mathematics which would still leave the physicists themselves thoroughly bamboozled, thereby keeping up the good work. It arrived in the form of **Quantum Electrodynamics**.

In the late 1940s, three physicists, **Shin'ichiro Tomonaga**, **Richard Feynman** and **Julian Schwinger**, working independently of each other, arrived at the same mathematical conclusions at the same time.

When interactions between light and matter are contemplated, the possibilities are infinite; and it was the 'infinities' in calculations which sent physicists loopy. Quantum electrodynamics, instead of banishing infinity from the sums, invited it to join the party. Miraculously it solved the problem.

QED is one of the most respected scientific theories ever. As Feynman was fond of pointing out, if you imagine that the calculations deal with the distance of London from New York, they would be accurate to within the breadth of one human hair.

Such accuracy is a good thing to invoke; by 1950 quantum mechanics had become what **Paul W. Davies** called "...the most powerful theory known to mankind" and what the *New Scientist* termed "...the most successful and wide-ranging theory devised by human ingenuity".

There were those, however, who felt that the intricate edifice stood on quicksand. Einstein, whose work with light and electrons had opened the curtains on the whole quantomime, wavered between calling quantum mechanics 'incomplete' and declaring its ideas to be "the system of delusions of an exceedingly intelligent paranoiac, concocted of incoherent elements of thought".

Obviously you are in the presence of a curious blend of derailment and reassurance. There will always be those who feel that confusion is entirely appropriate in God's territory; so you are absolved from all obligations to understand any of it, and need merely pick and choose from such descriptions as you deem worthy of exploitation.

But the incomprehensibility of quantum physics is a secret which you should guard with your life.

Matter

The greatest potential for paralysing another human being, far better than tickling, or getting a submission with a step-over leg lock, is to introduce a novice to the basic structure of matter.

All matter is made of molecules, you say, and these can be further divided into atoms. Explain that the atom was accepted as the basic building block from 430 BC until 1897, so it is a 20th century fashion to talk blithely of 'sub' atomic particles.

The kernel or core of the atom, the **nucleus**, is one ten-thousandth the size of the entire atom, and is itself composed of particles (**protons** and **neutrons**) bound together by a strong nuclear force. Protons carry a positive electrical charge. Neutrons are electrically neutral.

Whizzing round this positive-and-neutral nucleus are particles of negative electricity called **electrons**.

Try to imagine, you say, a 'shell' effect around the nucleus (like the illusion of solidity produced by a whirring top) which is formed by these electrons while zooming at a speed the equivalent of four times round the earth in one second. In fact, of course, they are whirring around a nucleus which is itself only one ten-thousandth the size of an atom.

To get an easier idea of scale, ask people to picture the dome of St. Peter's basilica in Rome. The electron shell would be the dome itself, and within the space of the dome – more or less at the centre – the nucleus would be the size of a grain of salt. If you remember to add that within the outer shell there may be layers and layers of inner ones, like skins, then they will really know their onions.

It is worth emphasising that electrons are infinitesimal points or wisps, and not planet-like spheres; they do not orbit, they flit about like supercharged moths. Even the fact that particles 'spin' at the same time as they zoom and jiggle should not lead anyone to think of the earth's rotation. Apart from anything else, particle spin occurs in two directions, 'up' and 'down' (a typically idiosyncratic physicists' expression for clockwise and anti-clockwise). It was **Wolfgang**

Pauli who showed that a single particle will insist on being alone and spinning in its own way in its own energy state. This is called the Exclusion Principle. If another particle joined it, it would have to have opposite spin. A third would make the others feel exceptionally crowded and would have to go off on its own or find another one with which to have opposite spin. You should limit the ways in which you let anyone say it is just like life.

The balancing of electrical charges at the subatomic level, is carried on throughout the universe: all the negative charges balance out all the positive charges. To some this may suggest that the universe all adds up to nothing. To others it presents another example of cosmic balance.

You may care to enlarge a bit here by stating that protons are 1,836 times bigger than electrons. Scientists tend to state this precise statistic and then say that neutrons are 'slightly' bigger than protons without saying how big slightly is. You have little choice but to follow suit, or to forget these particular facts at once.

But this isn't all. Not by a long shot.

Protons and neutrons are themselves composed of still smaller particles. Yes, 'fraid so. They are the notorious **quarks**.

Protons and neutrons and the quarks within them cluster together to form the nucleus of the atom, so some nuclei are very bobbly, like little clumps of caviar, or like raspberries. Just a touch more difficult to bite into.

Billions and billions and billions of atoms make up every tangible and visible object, so considerable mileage may be had from pointing to something and stating that it is just nuclei, space, whirr-effects and spin which create the solidity that we see and touch. What may look still and block-like is, in fact, motion. This is why holists speak of everything as being an illusion. (You can refute this with a well-aimed kick.)

Erwin Schrödinger suggested the entire universe and everything in it, the yaks and yams and stars and suns, the whole caboodle, was created by a single particle on a journey of incredible rapidity and wild complexity. Schrödinger was known for taking to his bed before and after such pronouncements; he was in a place to which any of us could so easily go.

Quarks and The Standard Model

Questions about subatomic particles may arise, indeed you yourself may wish to ask them, along the lines of: "Just a minute, no-one can actually even see these little thingie-whatsits, can they? Or "How do they know?" Or "Are drugs required?"

After all, subatomic particles are, well, sub-atomic: minuscule point-like entities (variously described as tendencies, or even dreams) which leave traces among jets of particles when there are collisions in the quantum realm. In short, they are not seen, they are detected. No-one had actually detected an atom until the 1980s with the invention of the scanning tunnelling microscope which can magnify what it is looking at 100 million times.

You could proffer the information that down the decades, experimental physicists have used photo-multipliers, and have developed voltage doublers, van der Graaff generators, cyclotrons, Geiger counters, cloud chambers and even massive tunnels miles wide in which particles have been smashed, accelerated, collided, traced, discovered, annihilated, created, presumed or given silly names. But the impression you should be giving is that you live a life of the mind, not of toil at the coal face. You should not deign to spend more than 20 seconds on any of this, nor the fact that you really did mean to say 'scanning tunnelling' microscope.

For many years the so-called 'Standard Model' – the orthodox view of what happens at the subatomic level – was composed of quarks and **leptons**. NB: You cannot say 'quarks and electrons' because electrons belong to a wider genre of particle called leptons, so while electrons are always leptons not all leptons are electrons. Fortunately no-one is ever going to ask you to name the others because this is about as much as any brain can take.

It would have been helpful if you could have just left it at that, especially since the nucleus is packed with six different kinds of quark and six varieties of lepton dancing together. That's it, you could have said. The basis of all matter. Excellent. So simple.

Except it is now suspected that quarks are made up of even smaller particles, at present called sub-quarks, and that the sub-divisions go sideways as compulsively as they go downwards. So that there may be not just six quarks and six leptons but a menagerie within the Particle Zoo.

Recent dramatic particle smashings have revealed the possibility of a strange basic entity, a 'leptoquark' which might simplify matters. To a physicist a leptoquark is a 'chalk-cheese' or a 'sow's ear silk purse', but it has been ages since anyone expected Nature to do what she's told.

Quarks were originally given 'flavours' whereby, for the purposes of calculations and the kind of work they seemed to do, they could be told apart. These flavours are 'up', 'down', 'strange' (originally called 'sideways' but it became even more odd), 'charm', 'bottom' and 'top'. (The latter two were originally called 'beauty' and 'truth'.) You should thank **Murray Gell-Mann** for these quirky quark names; they may be pretentious but they are a tad friendlier than mathematical symbols.

When **quantum chromodynamics** was developed in 1977, quarks were given 'colours', which have less to do with tone than flavours had to do with taste. You can get each flavour of quark in red, green or blue, and antiquarks (see below, or in another universe) in minus-red, minus-green and minus-blue.

In all fairness, physicists themselves find it difficult to hold on to their sanity in this reductionist's heaven, and have expressed disquiet about the Standard Model. Its calculations have proved successful, but as Einstein might have said, "If you were in charge, is that the way you would construct the universe?"

Antimatter

All of quantum mechanics has an Alice In Wonderland feeling about it. And **antimatter** is the looking-glass world. Indeed, the existence of antimatter is one of the arguments put forward for the existence of parallel universes. At least, in this universe it is.

Antimatter is matter composed entirely of antiparticles – elementary particles which have the same mass as given particles, but opposite electrical or magnetic properties. Protons are mirrored by **antiprotons**; neutrons by **antineutrons**, and electrons, which are negatively charged, by **positrons**.

The existence of antimatter was first suggested by Dirac. He was not the only person to notice during the Great Depression that there was an awful lot of negative energy around, but he had a novel way of looking at it. He suggested that it was simply an accident that the earth is made with a predominance of matter over antimatter, and that for some stars it may be the other way around. (You couldn't tell at a distance.)

Antimatter features in the notion of supersymmetry because it is believed that each and every particle has a twin in the looking-glass world, which – apart from its spin – is identical. All that can be said now about such an idea is 'Possibly'.

Physicists are very fond of making particles and antiparticles collide, and it is a treat for them when, as a result, particles and antiparticles 'annihilate to energy'. Still more magical is that nano-moment when out of pure energy, particles and antiparticles are created. It's enough to make them feel like Steven Spielberg.

Yet without the help of physicists this dance of creation and destruction is taking place all the time – in people, fax machines, walls, toothpaste, rubbish bins, and jelly beans. Nothing could exist if it didn't.

The Quantum Leap

When you speak of the whirring electrons inside every object, you need to bear in mind that this is what is happening at room temperature. It may be a hectic flurry to you, but to an electron it's repose. Heat them, though, and they really get going. The hotter they get the more excited they become, and when they reach a certain pitch of excitement and heat they 'radiate'.

With radiation the electron actually jumps or is boosted through different layers around the nucleus inside the atom. This is what Bohr called the 'quantum jump', an expression which gave rise to the 'quantum leap'.

You could do some interesting dinner-party dribbling with this one. Firstly, what Bohr spotted were jumps down rather than up. (If you need to know, an electron jumps downwards when radiation is emitted, and upwards when radiation is absorbed.)

Secondly, the term has come to mean a huge or sudden advance. Quantum leaps do seem to disobey a number of traditional laws of physics; and, relative to the size of the electron, they can involve impressive energy and range. But a quantum leap is a seriously small event.

Billions of such leaps are involved in every lightbulb, candle, barbecue, soldering iron, combustion engine, cigarette, their respective electrons all receiving and transmitting like one vast telephone exchange – with light, heat, energy and radiation instead of voices on the line.

Bohr and Heisenberg described the behaviour of these minute interacting particles as 'crazy', and Feynman – the physicists' physicist – with his customary American charm, called their antics 'nutty'. Today's physicists are reviving the word 'weird'.

It must not be forgotten that scientists are very restrained in their use of language. Nor must it be forgotten that many scientists have been unable to come to terms with the behaviour of subatomic particles. Nothing for them in two and a half millennia has been as shocking.

THE SHOCKS

In order to captivate any gathering, some appreciation is needed of the shocks received by classical scientists striding purposefully out of the 19th century.

Such was the respect accorded at that time to anything scientific, that whatever was unscientific was looked at as if it was a schnitzel that no-one had ordered. Indeed, as the 20th century began, it was believed that all the problems concerning nature and the physical world had been solved by science. Those minor problems which remained would be solved before Wednesday week, at the latest.

Wednesday week came and went. So did many very clever people, and within one quarter of a century all notions of space, time, gravity, energy and matter were changed. It was discovered that things bent, warped, scrunched and dawdled. And this was decades before rock 'n roll. It was shown that if you went faster than the speed of light you would go back in time, and that the faster you travel, the more distorted your body would become. Furthermore, the faster you travel, the slower your brain would run. Well, planes, trains and motorways are full of examples of that.

Perhaps most alarmingly, it was established that if your parents zoomed off at a speed getting on for the speed of light and stayed away for 40 Earth years, they would come back younger than you. And you could get your own back. And so could they.

All these things were accepted by Einstein – after all, he proposed them. But what he could not absorb were the shocks of quantum physics. When asked why, he said: "A good joke should never be repeated twice." (Einstein's strong suits did not include grammar. Nor, as it happens, maths.)

Within the Quantum realm the rules of nature are turned upside down. Quantum theory challenges, at the very least, notions of Causality, Predictability, Reversibility, Continuity, Locality, Order, Separateness (sometimes called Isolability), Clear Definition, Accurate and Informative Measurement, Objectivity, Either/or Thinking and, most controversially, Certainty. It threatens notions of solidity and substance, and suggests that there are absolutely no absolutes.

Causality

Causality is one of the most important pillars of science. Everyone knows that if you freeze water you will get ice and if you heat it, it will evaporate. But in quantum theory events seem not to have such well-defined causes and effects.

It all boils down to the Holist/Reductionist argument. Coming down on either side is dicey. Holists are convinced that an electron can jump between orbits for no discernible reason, or that a subatomic particle can come into being or disintegrate altogether without any cause. In other words, holists insist that particles do the loathsomely worst thing anybody or anything can do to a scientist: they behave spontaneously.

Reductionists share Einstein's lifelong distaste for the incorporation into science of anything 'acausal', random or 'stochastic' – a good word to use if you want to say 'subject to guesswork, or conjecture'. He said, "I cannot believe that God would play dice with the universe".

If you look at a glass of mineral water, Coke, or best of all champagne, you might begin to see the problem.

To analyse or anticipate the behaviour of any individual bubble is fiendishly difficult, even if it is momentarily motionless, let alone when it suddenly decides to jiggle around, disappear, or shoot off in funny directions.

And quantum physicists have set themselves a task far tougher than relating a bubble to a drink, or even an atom to a bubble. Yet reductionists say they have an answer for all that fizz, zoom and kapow!

The real kapow! lies in the fact that their methods deal with particles collectively, via the distribution of probabilities within what they call a macro system, i.e. the whole picture. In other words their solution is holistic.

Meanwhile holists say that it's no good calculating collectively: what about the individual particle? In other words they are thinking like reductionists.

You only need to point to this transposition of views to be protected in any discussion about the stochastic, random, or spontaneous nature of subatomic particles.

You will even find a number of physicists, who dismiss spontaneity – just like that.

For them, the central contentious issue when it comes to causality is Time. Is causality linked to time as profoundly as the word 'consequence' is related to 'sequence'?

Can an effect precede its cause? It's a question which, of course, arose from the next sections.

Predictability

The remarks about causality also apply to the issue of predictability. Anyone could have seen that coming.

Reversibility

Reversibility may be understood by watching a video of some movement in nature in what is loosely called real life; a person or an animal running, or swimming, or a speeded-up film of a flower blooming.

When such a scene is shown backwards it always looks odd, not life-like, even funny, because of what is known as the irreversibility of Nature. Natural processes take place in time, and they cannot be 'run backwards'.

However a video or film of a subatomic particle in motion could run for a century or indeed, conceivably, for several millennia, without looking odd if it were being run backwards. The viewer might have clues, identifiable events perhaps, or a peek at the label, but no-one coming fresh to that video as it played would be able to identify either the forward or backward motion of it. In short, subatomic particles are somehow outside Time – occupying a gap in the very nature they compose.

Some have been in existence since the universe began. Others have been created as recently as a second ago, and might have disintegrated a relatively long time before that second was over. And there is a strong possibility that there are subatomic particles whizzing about here and now which were created in the future.

It is comforting to realise that even the greatest geniuses when contemplating these phenomena are 'up against the limits of what can be known'. This is a handy expression to use on its own or in conjunction with 'up against the barriers of language'.

The latter ploy is better if you want to imply that you do have the knowledge but it is the language that is letting you down. The former is excellent if you need to be humble. And brief.

Continuity

If you were to arrive in the quantum realm, what you would first notice is the lighting. Strobe, unfortunately. It is an irritating flicker, flashing much more rapidly than in any nightclub. And the particles, like dancers in a nightclub lit by a strobe, are known to be 'here', then to be 'there', but no-one knows what happens in between. Their progress is discontinuous.

In classical science it is possible to 'trace a path', 'record a progression', 'monitor a process'; but the lives of subatomic particles are mysterious flickerings, and the bursts of knowing where they are, and what they are doing, are shorter than the gaps of disappearance and unknowability. Like teenagers, really. And like teenagers, they are always on the threshold of becoming – at the impossibly awkward stage of not knowing what they are.

The Particle/Wave Problem

In the quantum universe there is an essential and baffling fact: subatomic particles can also be waves. This you should speak of as the 'wave/particle duality': you should say it is the central mystery of quantum physics.

For a reductionist the mystery concerns the so-called 'quantum amplitude': a particle is of a different size from a wave. It ought not to be one and the same. To a holist there is no mystery; it is glorious and natural that nature's tiniest possible entity – a particle – should have connecting ripples throughout the universe.

With particles and waves you see science in a very bad light, and indeed it was in the course of a dispute about light that the major surprises suddenly appeared. Newton had stated that light was composed of particles. He called them 'corpuscles' and expected everyone to agree with him because he was a genius, and they did until in 1801 a certain **Thomas Young** proved that light was made up of waves.

Throughout the 19th century, everyone was so sure of the wave nature of light that they concentrated on finding out about what was waving.

As with most romantic plots, electricity and magnetism, which had started out not being suited to each other, were brought together and engaged. **Faraday** did a lot of the basic matchmaking, then **Maxwell** showed that 'electromagnetism' travelled at the speed of light – and he married them. Light *is* electromagnetic radiation. It's the kind of fact that gives the mind a lemony taste and then becomes pleasingly forgettable.

While light is always electromagnetic radiation, not all electromagnetic radiation is light. The light that can be seen makes up only a small part of the kinds of electromagnetic rays there are in the universe, which include X-rays, gamma rays, microwaves and even radio waves.

In the sorry saga so far, Newton had said that light was definitely made up of particles and Young had said light was definitely made of waves. Maxwell, Hertz and many others agreed: yes, light was made up of waves. That was that then. Until Einstein working with the photoelectric effect stated *'Nein'*, light is definitely made up of particles.

He didn't deny light's wave nature; but he could no

longer deny light's particle nature either. Which was why the pioneers of quantum physics had to have another look at an old experiment – the one with the two slits – which you should refer to as the **Double-Slit Experiment**.

Quote Feynman who said that all discussions about quantum mechanics will eventually come back to the experiment 'with the two holes'. The apparatus consisted of a light, a screen into which two very thin vertical slits were cut, and a wall. In Young's test the light had clearly lapped and undulated its way through and away from the slits, because there were blocks of light and shade on the wall. Particles of light would have flown like bullets, leaving no shading, just two clear slits.

These days, the very same test shows that light is made up of particles *and* waves. If you use a particle detector and test for particles you get particles. If you use a wave detector you get waves. Somehow, and no-one knows how, it is the test itself which gives you the result. It is utterly alien to classical physics.

Eddington once said that on Mondays, Wednesdays and Fridays an electron is a particle and on Tuesdays and Thursdays it's a wave. (He didn't work at weekends.)

And that's the way it is this Sunday lunchtime.

Accurate and Informative Measurement

It is important to make it clear, in a magnanimous and sage-like way, that the measurement of subatomic particles is phenomenally precise. Scientists, especially reductionists, cannot hear that often enough.

The problem is that scientists need two or more

measurements to work together, so that other conclusions may be drawn. If it is known that a plane is travelling at 500 miles an hour and that Rome is 1,000 miles away, then the plane should take two hours to arrive.

Scientific data should do more than one job, and in classical science position and momentum (by which physicists mean velocity times mass, rather than speed) go together like the Blues Brothers, Laurel and Hardy, or trains and dorks with mobile phones.

The shock when it came to quantum physics was that it was possible to confirm the position of a particle, but not its momentum: the particle was too small, too inaccessible, and too capable of deflection – as always – by the instrument, and the light used to trace it. All concentration was thus focused on finding the momentum. In fact, the momentum was established – with equal accuracy – but by then the particle had zoomed past and its position could not be pinpointed with anything like the same accuracy.

In any discussion about this conundrum you really need do nothing more than utter the three words **"Heisenberg's Uncertainty Principle"**. It is possible to be certain about one measurement, but then another would become uncertain.

Inevitably, the Uncertainty Principle is not principally known for what it has to say about uncertainty. Nor for its remarks about measurements.

The really terrifying bit of uncertainty for scientists, and indeed all of Western society, concerns objectivity.

Objectivity

If you ever get the urge to challenge classical science, use Heisenberg's statement: "The act of observation affects that which is being observed." And then watch.

In science any test must work independently of the person conducting it. The worst put-down one can have from a scientist is "Now you're being merely subjective" – i.e. influenced by individual feelings or opinions, and therefore irrational. Only objectivity, the collective subjective opinions of like-minded scientists, can be trusted. Like a prop forward, objectivity is butch and dependable. It won't cry.

It is from objective truth that the laws of nature are derived. And these laws hold good whether or not someone is checking on them. With objectivity, the observer cannot affect the experiment; the experiment and the observation of it are separate in every way.

But in the quantum realm it is hard enough to keep the experimental apparatus separate from the experiment itself; there is the added problem of the physicist always getting the result being tested for. The human being is profoundly involved.

This is where Schrödinger came in with his cat.

Schrödinger's Cat

Schrödinger's 'thought experiment', proposed that a cat be placed in a sealed box containing a small flask of cyanide gas which might (or might not) be shattered by a hammer that might (or might not) be activated by an emission from a lump of (potentially) radioactive material – and the experiment be run long enough to give the cat a 50/50 chance of survival.

The cat represents two possibilities, or probabilities.

It could be alive or dead and continues to be in that state until – well, until the scientist takes a look. Until then, it is in an absurd and, it has to be said, unscientific state of live-deadness.

The point is that Observation Kills The Cat.

This story should be dispensed with very quickly. Apart from anything else, cat-lovers never take in the details, they just want to get the cat out of the box and cuddle it. However, for some, what is important is that as the act of observation takes place, reality forks and two entire universes are simultaneously brought into being, one in which the cat is alive and scratching and another in which the cat is dead, deceased and gone to its maker.

Indeed, many boffins go further and declare that the two possibilities represent two separate universes and that, in each of those universes, there may be other twinned possibilities, and other universes. For them, reality doesn't just fork, it fans. This gives a potentially endless number of realities and is called the 'many worlds' idea, though you could strike more chords in speaking of 'parallel universes'. All's fair – providing you remember that ever-absolving word 'possibly'.

Schrödinger himself went back and forth wondering whether or not he agreed with quantum theory. He often took to his bed. Equally often, Niels Bohr would follow him into the bedroom and talk at him some more to try to convince him that quantum events were everything and headaches were nothing. It is much regretted that the poor, kind man had such frequent headaches, colds and 'flu. In a poetically just universe he would have been allergic to cats and would have swelled up and popped. (Then the rest of us could have Observed it.)

Locality

People have been known to scold with the words, "I can't be in two places at the same time." Yet it seems as if subatomic particles can. People say it automatically, like "How are you", or "This is so civilised" as they pour the tea, or "I'm sorry, he's in a meeting"; they just trot it out, "Of course, subatomic particles can be in two places at once."

Nothing goes to the heart of the differences between the reductionists and the holists like the question of locality. As far as the reductionists are concerned, classical science has managed more or less to accept and understand that particles of light do become waves of light and vice versa when they have to negotiate, say, the pupils of eyes, or telescope lenses or camera shutters.

And classical scientists never thought to question the fact that such events happened 'locally', at the cricket ground, or Julia's wedding, or through Brian's window; the light negotiated a small hole, or thick pieces of glass, or a thick piece of glass and a hole, and moved on.

For the holists, there is the knowledge that Heisenberg and Bohr, when they developed quantum theory, began to be aware of something vaguely mystical about it all. Bohr incorporated the Taoist symbol of oneness into his coat-of-arms. Cult books about quantum physics published in the 1980s created considerable excitement among Buddhists, vegetarians and ageing hippies: "I mean, don't look for why, man, don't look back, don't explain, know what I'm saying? It just happens, man. Spontaneously. God – why won't my van start? What are you doing next summer?"

This **weirdness**, or **non-locality** is still controversial, and therefore a winner for you. The focus of the dispute concerns interactions between particles that are in different localities. Einstein said no such interactions over any distance could take place faster than the speed of light. Nevertheless, even in his day it seemed as if interactions between particles did take place suspiciously simultaneously. Bohr's argument was that separated particles were still part of a single totality, even if one was on Earth and the other on another planet.

To challenge this, Einstein and two younger colleagues, Podolsky and Rosen, dreamed up a thought experiment, the **EPR Experiment**.

E., P. and R. believed they could show quantum mechanics to be 'incomplete'. They suggested an imaginary collision between two particles, which then fly off separately to some distance from each other. Then by knowing the position of the smash and by probing one of the particles immediately afterwards, all the data about position and momentum concerning both particles could be calculated. Of course, this examination would have to be extremely quick. It would have to happen in a flicker of time. But it would be worth it to contradict Heisenberg, and indeed, to nullify all the shocks wrought by quantum mechanics and save classical science.

There would only be one way in which the E.P.R. experiment could ever be wrong: if the messages between particles travelled faster than the speed of light (i.e. that they were 'superluminal'). In that event, it would not be possible to probe one particle without simultaneously probing the other. Such a notion was worse than disturbing to Einstein; it made scientists look like fairground psychics, gamblers or dabblers in voodoo. He called it 'spooky and absurd'.

Einstein died. Bohr died. Then, in 1964 **John Bell** at CERN in Switzerland published some challenging calculations. Bell had actually begun work in the hope of proving Einstein and his young assistants right, but his Theorem opened the door to Einstein's worst nightmare. Repeated experiments confirmed that once two particles have had any interaction they do somehow remain linked as parts of the same indivisible system. Separated particles seem to be as connected as two ends of the same rod.

You can expect discussions about the connection between twins, the collective unconscious, television ratings – any number of subjects can be swept into this gigantic dustpan.

Try to remain ambivalent. Make the point that Bell's Theorem and the experiments provide indirect proof; that is to say, the powerful rod-like connection has to be there because Bell's sums say it is. Predicting an election result because of opinion polls isn't quite the same thing.

Be gentle as you cite Bell. However smug it feels to have proved Einstein wrong, this should not show. Firstly, it would be unseemly. Secondly, you would be hated for it – many people love him as if he were their puppy. Thirdly, there are certain objections which he raised that have still not been answered. What if, tomorrow, new particles are found which are separate, predictable, local, and he were proved to be right?

Order

It is worth noting, for conversations in which the listeners do not mind getting technical, that all electrons are identical with each other, all protons are identical with each other, and all neutrons are identical with each other. In other words, three basic entities, by combining to form atoms, which combine again to form molecules, combine and combine again to build everything in the universe from bratwurst to battleships, from waffles to warthogs, from neck braces to New York.

But it will also by now be clear that in every cubic millimetre of space there are billions and billions of subatomic particles which are capable of behaving at times in astonishing and unpredictable ways for no rhyme or reason. These repeatedly, spontaneously and 'nuttily' change their physical qualities, their whereabouts, their energies and their flight paths and, small as they are, seem to have the entire universe in tow. It is quite difficult to put them in neat piles, label them and file them away.

Clear Definition

The point never stops being made, unfortunately, that particles which can clearly be defined as particles can also be found to have none of the qualities which are recognised as particle qualities; at such times they have all the qualities of waves and must be defined as waves. This is not a new idea, and yet the shock waves from it still do not seem to have subsided.

It challenges classical science, because clarity of definition is a cornerstone of science. It is almost the definition of it.

Separateness

Particles are often defined as 'discrete'. This is a good word to use. 'Discrete' means 'individually distinct', 'separate'. Holists believe that each particle tugs at and is tugged by the universe. Reductionists cannot quite budge from their stand of separateness and splendid isolation.

Whether you support one or the other (or neither) you need to concede that it is possible to study and conduct experiments with individual particles. Indeed, the ability to send a single electron whizzing through silicon is the very basis of the entire computer industry. (This fact should get good reactions. If people are receptive, it is good drama to add that transistor radios, digital systems in television and sound systems, and even – pointing around the table at a sequence of wrists – digital watches, could not have been developed without quantum physics. Or you may prefer doing it the other way around, making computers the big finish.)

But, when trying to study subatomic particles, it is impossible to subject them to traditional, classical scientific examination, because particles relate to each other and to equipment itself with such determination. It is sometimes stated that these particles are not experimentally 'isolable'.

It might seem obvious that this should be so, but Einstein didn't like it. He didn't like it at all.

The Three Breakfasts

In 1927, during the Solvay conference in Brussels, held in the Metropole Hotel, Bohr and Einstein conducted a debate which you might choose to call the greatest intellectual encounter since Newton's famous clash with Leibnitz. Using the word 'famous' when people are unlikely to have heard of something is an unkind, but delicious, ploy. It helps the story along if you mention that they conducted their arguments outside the business of the conference, over three consecutive breakfasts.

Einstein was in effect defending the Newtonian universe against all those Shocks with which you are now familiar. Suggestions that subatomic particles would be affected by the very equipment with which they were examined went against Einstein's deepest convictions.

At the second breakfast he made up a 'thought experiment' which he was sure would prove that a single particle could be released without affecting the equipment at all. As he finished his croissant, it looked as if Bohr had lost.

But at the third breakfast, Bohr proved Einstein incorrect. Not only that, but Bohr was able to show that the equipment would have to be affected because of Einstein's own great Theory of Relativity.

Scientists might suggest Bohr was convincing the great man that quantum theory is not inconsistent with the Theory of Relativity. Bluffers may want to be fair to Einstein, and allow that one through. Hardhearts may wish to push the point that, for once, Einstein was wrong.

Nevertheless, what Niels Bohr was working on at the time of his death in 1962 was a diagram of that thought experiment proposed 35 years earlier in Belgium by Einstein. He was probably just reassuring himself one last time.

Either/Or Thinking

All the new shocks and questions have not shaken the fundamental difference between holists and reductionists. Reductionists say 'either/or', holists say 'both/and'. Both can be either right or wrong, or indeed, both right and wrong. This applies to both 'either/or' and/or 'both/and'.

Certainty

Obviously if definitions, accuracy and predictability are at risk, Certainty will be in trouble too. Note:

1. The Uncertainty Principle refers to the phenomenon of uncertainty with regard to the position and momentum of a particle. It is not saying that nothing is to be trusted anywhere any more.

2. Classical science is still for the most part accurate and certain. You need to say this, even if you do feel moved to add that reductionists need to accept some doubt, some spontaneity, some life.

3. All possible or probable jokes about the Uncertainty Principle have been made and made and made again. You can be sure of that.

COOKING YOUR OWN GUT

In the Quantum Universe the end of physics will be found. And thereby, the solution to everything.

While it is clear that there should be a **Grand Unifying Theory** (GUT) and there have been any number of attempts to prove that there is, all such comprehensive proposals remain satisfyingly loose.

There is no reason why you should not join the legions of super-ambitious physicists in the search for this **Theory of Everything** (TOE) which it was Einstein's dream to find. (In Einstein's day, a GUT or TOE would have been called an UFT, for **Unified Field Theory**, because it had been discovered that permeating the physical universe there are four kinds of 'fields', and that, sportingly, they allow four 'forces' to frolic there.)

In the 1950s Heisenberg said he and his friend Wolfgang Pauli had just about got there. Pauli, however, drew a cartoon of an artist, brush in hand in front of a canvas on which there were a few lines and squiggles. The caption read: 'I can paint like Titian. All that's left are a few technical details'.

This should inspire you to cook your own GUT. You are ready to do so.

First catch your particles of matter. These can be difficult to clean and separate, so just cram them in the pot and persuade as many as you can to stay.

Then get hold of your messenger particles. Be warned: if you thought particles of matter were unwieldy, wait until you try dealing with these.

You will find them in the forces of nature which you have been marinating at least since the Big Bang. In ascending order of their powers of attraction or repulsion the forces are:

1. **gravity**,
2. the **weak nuclear force**,
3. **electromagnetism**, and
4. the strong nuclear force, which you should always refer to as 'the **strong force**' – the immense tug at the core of atoms. Nothing in the universe binds like it. Its particles are called **gluons**. (It is the breaking of this binding which makes nuclear weapons wreak havoc; and the harnessing of it which creates nuclear energy.)

By the time you have stewed, minced, mashed, stirred, whisked, and added a pinch of salt, you would hope that they would all be reduced – after straining – to a single force.

But be prepared for disappointment. A GUT has never been achieved before (nor has it been Grand or Unified) for the simple but persistent reason that gravity has been left out and even the other three have not yet been united.

Stepping Towards a TOE

The major stumbling block is that subatomic particles have always seemed free of gravity, flying about with as much abandon as international businessmen. More to the point, the maths makes utter nonsense.

Einstein laboured on this problem for decades trying to unite gravity with the other three forces, so that quantum physics could co-exist with his General Theory of Relativity and subatomic particles would make sense within the universe as a whole.

Uniting the weak nuclear force with electromagnetism was only achieved some 20 years after his death.

Since electromagnetism has a vastly longer range, operates more symmetrically, and is ten billion times stronger than the weak nuclear force, you should regard the discovery that they are the same fundamental force as spectacular, and call the theory which proves it the **'electroweak theory'**.

The men who came up with this, Weinberg, Glashow, and Salam, did so independently of each other but at exactly the same time which is the kind of thing that turns holists all a-quiver. The particles which carry this combined force are called, inventively, W and Z+ and Z- which excites reductionists and no-one else.

So much for what has been done. For the rest, the belief is that the more billions that are spent on them, the more answers they will find. A tunnel in the Texan desert, for instance, cost the Americans mega-dollars (think billions, think double that) before it was stopped, and anything they do with the tunnel at CERN doesn't come cheap. You can expect results from some tunnel somewhere any millennium now.

What are they looking for? For one thing, **Higgs' boson** (named after Peter Higgs, British physicist). Electroweak particles (W & Z to their friends) are puzzlingly heavy and it seems as if Higgs' bosons feed them mass, which is impressively Creative. This is why the Higgs' boson was for a short while called 'the God particle'. You should say that Higgs' boson will not be the whole answer, but just a step along the way, a missing piece in the giant jigsaw. Possibly.

Strange, new, undiscovered entities such as Higgs' bosons or leptoquarks have brought with them the possibility of another force – a **fifth force**, as if the clumsy four-fingered cosmos had suddenly gained a skilful and adaptable thumb.

Nobody has given this idea a lift yet, but in the

desperate rush to throw money at any angle of a TOE, the $500 million investigation of one of the four known forces, gravity, was begun in 1995. The search for gravity waves and gravity particles has so far resulted only in a name for the latter – **gravitons** (what else?).

Of the piles of other TOEs and GUTs lying around the place, there is one other you might want to single out, if only because many top physicists are dedicating their lives to it. This involves the **neutrino**.

Enrico Fermi suggested this word (Italian for 'little neutral one') for particles which have no electrical charge, are subject to the weak nuclear force and are capable of passing through a block of lead which stretches from here to Jupiter. What role they may play in the picture of the universe as a whole is still up for grabs, or not, as the case may be. Pauli offered a bottle of champagne for anyone who could prove that they existed. He paid up.

The conviction remains that the day is not far off when the world will finally have an all-encompassing Theory of Everything. This is why **Superstrings** and **M-Theory** seem so attractive.

Superstrings and M-Theory

String Theory and M-Theory bring physicists the closest most of them will ever get to being drunk. As with all GUTs, TOEs and UFTs they will answer questions about the beginning of time, the origin of the universe, whether there are multi-dimensional universes, and the best way to make tea.

The idea is that the fruits of centuries of enthusiastic investigations, the totality of information about the universe will be encapsulated in a single equation;

and that if the right equipment was available, it could be seen that the ultimate fundamental entities are small vibrating strings or small membranes. (When you say 'small' know that these entities look up at an atom as an atom looks up at the entire solar system.)

Superstrings are believed to thrum like the strings of a musical instrument, and as they do so, a wide variety of energies emanate from them as in a highly complex chord. However, instead of notes, the strings produce streams of particles. So all of creation is somehow tuned into existence.

The man you should tip as the superstar of superstrings is the American, **Edward Witten** (whom journalists are already calling 'perhaps the greatest physicist of all time'). For Witten, it is a great scientific insight that among those particles these strings carry not just the strong force, but all forces. Therefore superstring theory doesn't just propose that gravity is included, it *requires* it.

M-Theory stands for Membrane Theory, Mother of all strings or Mystery, and proposes that strings coexist with membranes of various dimensions – as many as eleven, as it happens.

Imagine the kudos you could get when talking about these extra dimensions because people will believe you can actually perceive them. You can't. They exist in the sums and, the theory goes, in real life they exist (in effect) non-existently. The membranes whose dimensions exist as no dimension are called 'zero-brane'; the ones which have string-like dimensions are called 'one-brane', those with dimensions like a bubble are called 'two-brane', and so forth. No wonder its been suggested that if 'p' were chosen to represent the dimensions of these particles, they would all belong to the category p-brane.

FAMOUS PHYSICISTS

Whenever the conversation gets a bit sticky, it is well worth your while to have some snappy facts about the People Who Made the New Physics. It's especially good if no-one else has heard of them and you can blind everyone with scientists.

David Bohm (1917-1995)

Recording chats with the Dalai Lama and Krishnamurti, Bohm was one of the holiest of the holists. He was sure that there is an as yet invisible but underlying unity, an Implicate Order forming a background to all physical, psychological and spiritual experiences.

His calculations in the early 1950s led to later proofs of subatomic non-local quality, but then he lived a non-local life in the United States, Britain, South America and the Middle East, with interludes among the Blackfoot people in what reductionists call Canada.

Niels Bohr (1885-1962)

A family man, happily married for 50 years to Margrethe, with five children, Bohr led the group of physicists who developed one of the greatest theories of all time at his home. Heisenberg, Schrödinger, Fermi, and many others came to stay with him, wore down the floor at either end of the ping pong table, wrote letters home, munched on sandwiches and ran around with his laughing children.

Bohr, who was a member of the Danish underground during the Second World War, had a portentous walk with his friend and colleague Heisenberg in

1941. No-one knows exactly what was said, except that Bohr was extremely shocked and the friendship with Heisenberg ended there and then.

Possibly Heisenberg told Bohr of the Germans' plans for an atom bomb. Within weeks Bohr had contacted the Americans and was whisked to the U.S. As a direct result of Bohr's alarm, work on the atom bomb was stepped up at Los Alamos.

Bohr had an unfortunate way of speaking. Einstein's comment about Bohr's lack of ability to allow words to come out of his mouth properly was: "He utters his opinions like one perpetually groping, and never like one who believes he has the truth."

A fellow physicist once drew a cartoon of Bohr talking to a friend who is bound and gagged. Bohr is saying "Please, please may I get a word in?"

Max Born (1882-1970)

The Probability man. Born was already pretty nifty with the old physics before he made his vast contribution to quantum theory by sorting out the mathematical formulations and practicalities of the ideas of Heisenberg and Schrödinger. You should say that it is his technique – the Born Approximation – which is used by working physicists far more than the philosophical speculations of Heisenberg and Uncertainty.

You should be scandalised that Born did not win the Nobel Prize until he was 72 and even then had to share it. It's worth mentioning that Einstein's famous statement about quantum mechanics – "I cannot believe that God would play dice with the universe" was to his great friend Born (in a letter).

An extra little pearl to cast is that Born was Olivia Newton-John's grandfather.

Louis-Victor de Broglie (1892-1987)

De Broglie (pronounced de Broy) devised a simple mathematical relationship connecting the wave and particle properties of matter.

Einstein had proposed the possibility that there were such things as matter waves in 1905, and in fact it was Einstein's enthusiastic response to de Broglie's thesis that made his name – another instance of Einstein's support for quantum mechanics despite his loathing of the claims that were made for it.

De Broglie's great-great-grandfather was a French aristocrat who died on the guillotine.

Paul Adrien Maurice Dirac (1902-1984)

The word associated with Dirac is 'antimatter' which he was the first to suggest might exist. Of Swiss parentage and born in Bristol, Dirac was Feynman's hero. He makes many physicists quite misty-eyed because he believed that calculations and theories should be endowed with beauty.

His approach to mathematics was always elegant and simple. So was his approach to life. He hated wasting words. Once asked if he took sugar, he answered "Yes" and was surprised to be asked "How many lumps?" If he had wanted more than one lump, he as a mathematician would have specified the number.

Paul Ehrenfest (1890-1933)

A great friend of Einstein's, Ehrenfest had tears rolling down his cheeks during Einstein's debate with Bohr. He forcefully reminded Einstein not to be as rigidly against quantum mechanics as people had been against Relativity. Yet he himself called the quantum theorists '*Klugscheisser*' – clever excrement.

Many have felt overwhelmed by quantum mechanics, but Ehrenfest's was an extreme case. The letter he sent before he committed suicide began: 'My dear friends Bohr, Einstein [and others]. In recent years it has become ever more difficult for me to follow developments [in physics] with understanding. After trying, ever more enervated and torn, I have finally given up in DESPERATION.'

Albert Einstein (1879-1955)

Einstein was married twice (to Mileva and to his cousin, Elsa). His first child with Mileva was born before their marriage, and had to be given away because they were too poor to provide for it. His second son spent most of his life in a mental hospital.

Einstein worked alone. People who knew him well say that he was somehow always 'apart'. He said "I am not much good with people... I feel the insignificance of the individual and it makes me happy." It is not surprising he kept insisting that subatomic particles had to be thought of as separate entities – 'discrete'.

Einstein died in America, far from his place of birth. Reporters jostling around the nurse who had been the only person present when he died were told that, Yes, the great sage had spoken as he breathed his last. What did he say? "I'm sorry," said the nurse, "I don't speak German."

Enrico Fermi (1901-1954)

Born in Rome, Fermi was rare in that he actually did his own experimental dirty work, even in America. The

Fermi Award for innovation is still much coveted by physicists. Fermi built the first nuclear reactor. When he established the first chain reaction, a telegram was sent which said: 'The Italian navigator has entered the new world'.

Richard Feynman (1918-1988)

Feynman's work on light and matter perfected what is possibly the most accurate theory ever developed in science. His famous diagrams show quantum interaction. You should always stress that, though there is so much probability in quantum mechanics, there is also accuracy and precision.

Feynman (it rhymes with lineman) was a hugely popular figure, as much for his mathematical wizardry as for his personality. A brilliant lecturer, he enjoyed doing sums on paper napkins at nightclubs and was an accomplished picker of locks which, if you are dealing with atomic secrets and one of your best friends is the spy Klaus Fuchs (who gave American atomic secrets to the USSR), is a pretty risky business.

Feynman's diagrams were controversial because no-one knew how he managed to derive them, but few are willing to argue for long about a drawing that saves hundreds of pages of algebra.

Murray Gell-Mann (1929)

Gell-Mann had an office across the way from Feynman at Caltech University in the 1950s, and was judged the more urbane of the two. He sprinkled his scientific work with literary and classical references, and ordered particles into arrangements or families which he called the Eightfold Way in honour of the Buddha.

His most famous contribution is the 'quark' which he both discovered and named. Gell-Mann insisted that it be pronounced 'quork', to rhyme with pork, from a line in *Finnegan's Wake* by James Joyce: "Three quarks for Muster Mark." There are bigger mysteries in quantum physics for you to waste any time trying to figure that one out.

Stephen Hawking (1942)

Hawking (together with Sir Roger Penrose) proved that the beginning of the universe was a 'singularity', a mathematical point of infinite density, the explosion of which was the Big Bang. Hawking fathered a subject whereby, instead of wrestling to bring all the forces together, the quantum-mechanical implications for gravity alone are studied. It's called quantum cosmology.

Hawking loathes the idea of parallel universes. "When I hear the words Schrödinger's cat", he said, "I wish I were able to reach for my gun."

Werner Heisenberg (1901-1976)

Famous for his Uncertainty Principle (published when he was 26), Heisenberg is also notorious for having gained some advancement during the Hitler years in Germany. But no-one quite understands what his role was, because plans for a German atom bomb were abandoned by order of the Führer.

Heisenberg's musings about the observer and the observed have been exploited far beyond his intentions, and you should scoff at the cavalier use of the word Uncertainty which is not an excuse to avoid marriage or fill in tax returns.

Max Planck (1858-1947)

Famous for Planck's Constant – the energy of a light wave is always proportional to its frequency.

The mathematically-minded enjoy the strange relationship between Planck's Constant and Heisenberg's Uncertainty Principle. By multiplying together Heisenberg's two uncertainties you actually get Planck's constant. Truth is stranger than fiction.

Planck's constant is represented by h, which has the value of 6.63 times 10 to the power of minus 34 Js (joule second). It is always 6.63 x 10 to the power of minus 34 Js, and the energy of a photon is h – or 6.63 times 10 to the power of minus 34 Js multiplied by f, where f is the frequency of the wave. It should by now be clear why jokes about being thick as a Planck are not such a great idea.

Planck could never come to terms with quantum mechanics, of which he himself was a pioneer. A professor in Berlin for 60 years, his son was shot for attempting to assassinate Hitler.

Wolfgang Pauli (1900-1958)

At the age of 20 Pauli wrote a 200-page encyclopædia entry on the Theory of Relativity. It was his ideas that led to the discovery of the neutrino, but he is best known for his Exclusion Principle.

Pauli was excluded many times from bars for being pixillated. Even when sober, he had no trouble speaking his mind. To one student he said, "Ach, so young and already you are unknown"; to another, "That isn't even wrong". He even put Einstein down for not seeing the difference between mathematics and physics. It was to Pauli that Bohr made the legendary remark:

"What you suggest is crazy. But not crazy enough."

Pauli was also fascinated by subatomic particles and consciousness, collaborating for some time with psychologist Carl Jung, whose patient he was for a while. Their association may not have proved much, but it probably made them feel better.

Ernest Rutherford (1871-1937)

Rutherford, who sang 'Onward Christian Soldiers' loudly and out of tune all day and every day for the whole of his life, laid the groundwork for the development of nuclear physics by discovering the alpha particle, the nucleus and the proton. He was also wise enough to employ Niels Bohr, as well as Hans Geiger, who developed the Geiger counter as a result of sitting in the dark and totting up flashes of radiation.

Rutherford contributed to the atom bomb by yelling at a young Hungarian physicist called Leó Szilárd that such a thing would be impossible and throwing him out. Within 15 years Szilárd was a leader of the Manhattan Project, his chief encouragement an H.G. Wells novel called *The Shape of Things to Come*.

Erwin Schrödinger (1887-1961)

A brilliant physicist who stands out because of his extraordinary intellectual versatility, his contributions to science include an extremely useful wave equation, and a handy and profound book about quantum physics and genetic structure. Yet the poor man is remembered for that dratted cat.

THE IMPLICATIONS
What It All Means

Apart from the obvious fun that can be derived from believing impossible things for as much as half an hour a day with six of them before breakfast, quantum mechanics is a big part of everyday life. It deals with the most basic of basic stuff. It is behind every chemical reaction, every biological and medical miracle. It underlies all of existence in some way.

And therein lies the controversy. For the holists, what is so powerful is the discovery of these fundamental interrelationships which permeate the entire universe and make it a cohesive (w)hole. They feel that western civilization should finally resign its obsession with dividing, compartmentalising and separating – an obsession which has governed intellectual activity since the ancient Greeks, and especially since Aristotle whom they regard as the chief villain of the pieces.

They say it's time to stop dis-membering and begin re-membering. They hold divisive thinking responsible for all the world's splitting, clefting and rending; they say it's what causes revolutions, riots and wars.

The holists latch on to statements like that made by **John Wheeler**, a physics professor who worked with some of the greats, including Bohr. Wheeler stated: "Nothing is more important about quantum physics than this: it has destroyed the concept of the world as 'sitting out there'." Wheeler is very firm about no longer thinking of the universe as being for Observation. It is a universe which requires, demands and expects Participation.

Reductionists say that's too limiting.

The Cutting Edge

There are two fashionable topics at the cutting edge, or even the outer limits, whose popularity has mushroomed as a result of quantum mechanics. The first is mysticism, the second is consciousness.

1. Mysticism

To reductionists this means anything from wearing a saffron robe to having a fondness for breathing too deeply. To them there is no worse word.

Supreme caution must be exercised with mysticism in the presence of any member of the scientific establishment, unless you are tired of life.

2. Consciousness

The fact that it is considered to be no more than a possibility that consciousness is connected with phenomena in the quantum realm should not prevent you from pontificating about it. Sir Roger Penrose, sometime colleague of Stephen Hawking, goes farther. For him it's a 'definite possibility'.

Evidence for such a convinced 'maybe' lies in the swiftness with which the human brain makes choices. For instance, sifting all the visual alternatives as the eye focuses on a single printed word takes the brain less than a tenth of a second; a standard computer would have to start calculating before time and the universe began, so you wouldn't want to be stuck behind one of them on a roundabout.

But speculation about the act of observation takes the issue far beyond the ordinary miracle of the brain's astounding speed, beyond settling the fate of

Schrödinger's cat. It takes it towards affecting reality itself, and even into the clutches of Bishop Berkeley's assertion in the 18th century that Mind creates Matter.

With the double-slit experiment, are the particles and waves merely sensitive to screens and other equipment? Or are they picking up messages from the physicist's brain?

The mathematician John von Neumann, the biologist George Wald, the physicists David Bohm and Arthur Eddington have declared that the universe is mind-stuff, but it is difficult to get out of one's head for long enough to prove it, even in this day and age.

The softer option is to say that particles of matter and particles of mind may come into being together, but that in any case our self-awareness and what seems to be some kind of consciousness at the quantum level are in profound communication.

Consciousness is also non-local. No-one can say where the mind is, or how far the effects of thought can reach. The Anthropic Principles (both Strong and Weak – see Glossary) suggest that in man there are intellectual capacities which are there for a reason, that somehow human beings are obliged to help the universe through the next stage, although it is unclear whether this means giving it homework, or getting intoxicated with it and talking all night.

You can be supremely engaging and effective, without being too technical, by telling people about the messenger particles in the brain and how profoundly they reinforce the interconnectedness of everything.

Messenger particles – which by now you will be able to call 'bosons' without flinching – are the strongest, most basic forces of relationship in nature. If anything relates more fervently it hasn't yet been found. Indeed, within the brain, all the forces are active, including gravity, which is nothing less than the force that holds the entire universe together.

No wonder there are suggestions that the universe may be in some sense a Great Mind, and that a theory of everything will have to include a theory of consciousness. Perhaps superstrings are thought particles. And perhaps they know this is being supposed. Perhaps they supposed it first. Perhaps the supposing happened simultaneously. It's just a thought.

To Be Continued...

If questions in the quantum realm go to the root of everything and the answers which physicists seek will reveal the secret to Life, the Universe and Everything, then it is easy to see why, whenever you raise the subject, you are firing the starting gun for a conversational marathon.

The quantum universe has different rules about endings and beginnings. Many physicists believe that the Big Bang was a quantum event, the explosion of a single particle out of which the universe ballooned.

Whether the universe is contracting or expanding – crunching down to a single particle or creating from each particle a potential universe – there is nothing to stop you proposing that it will eventually evolve, dissolve and revolve into a quantum whisper.

GLOSSARY

Anthropic principle – The theory that the universe has evolved so that human beings could come along, and because of this they have a purpose, and their minds matter. The **strong** anthropic principle states that humans (Man/anthropos) matter a lot. The **weak** anthropic principle states that man matters quite a lot, especially if he/she understands the E.P.R. paradox.

Antimatter – Matter composed entirely of anti-particles.

Atom – The smallest unit of any material which still retains the characteristics of that material. Obviously if you go into the subatomic realm anything can happen and probably will. The word 'atom' comes from the Greek meaning 'that which cannot be cut or split'. Over-confidence is always perilous.

Big Bang – Generally-accepted notion that Time had a beginning, and an incredible amount of action during the credits.

Billion – Big number which should be bandied about with abandon. The British billion is traditionally a million million. In physics, the American billion is used: a thousand million. Piffling in either case.

Constant – A comforting predictable factor. The speed of light is a constant. So is the constant discovered by Planck. Useful in science and in life: buses are always late, they don't write songs like that any more, and money is never going to grow on trees.

Electromagnetism – A weak force. Ultra-violet rays from the sun are electromagnetic. So are the X-rays, microwaves in ovens and radio waves. Coming soon: cook your chicken while tuned in to Moscow.

Elementary particle – A particle without any internal structure which is therefore considered to be one of the basic building blocks of matter. In other words, the smallest things there are. So far.

Energy – Oomph, pzazz, vigour. In physics, as in life, it also means the 'capacity to do the work'.

Gluons – Mega-strong messenger particles which bind everything together in the nucleus. Far more binding than marriage vows, contracts or even eggs.

Gravity – The binding force of the universe, central to Relativity, left out of the quantum dance. If anyone doubts gravity's existence, point to the ageing process.

GUT – Grand Unifying Theory. A theory of everything which unites all the forces, the search for which goes on and on. On that, at least, physicists are united.

Laser – Light Amplification by Stimulated Emission of Radiation. The emission is stimulated from an excited source. (Don't even *think* of making jokes about this.)

Mechanics – Branch of applied maths dealing with motion and tendencies to motion. (Someone's got to do it.)

Molecule – Two or more atoms in a 'bound system'. There are 92 naturally occurring substances (called elements) and another 17 man-made ones (so far), the atoms of which can combine to form molecules which ultimately form everything from guacamole to Guatemala.

Nucleus – The dense core of an atom, made up of protons and neutrons. Nuclear energy and the nuclear bomb refer to this kind of nucleus. The nuclear family is so called because it too is a basic unit; to point out that families split, experience fallout or blow up is in very bad taste.

Particle – Small part (sometimes the least possible part) of something. It is the central mystery of quantum mechanics that a particle sometimes behaves like a **wave**. You are advised to hop on it and surf while the going's good.

Probability – Speculative factor essential to quantum theory and to any discussions about it. Bluffers need to know that a number of physicists prefer notions of probability to the use of the word 'uncertainty'. Gags to avoid: asking "How do you know?" "Are you sure?" or "Is that likely?"

Quantum – If you don't know now, you never will. But that doesn't mean you won't have a happy life.

Superstrings – Theory that minuscule vibrating strings will be, for certain, the smallest possible anythings in nature. A neat solution to bring the threads together? Reductionists are already sharpening their scissors.

Time – Still not found in the quantum realm. Once and future particles just go jiggling and jiggling and jiggling along.

Wave – A 'disturbance' like a ripple, a shiver or a vibration, but one which somehow happens to reach throughout the universe.

THE AUTHOR

Jack Klaff is a writer-performer whose workroom is living proof of the difficulty of finding Order in the universe. He has created a number of solo shows, as well as several works for radio and television on subjects as diverse as world affairs, psychoanalysis, physics, Kafka, myths, legends, and male/female relationships.

Amazingly physicists and science journalists have expressed delight with his theatrical interpretations of quantum mechanics, saying they find the work oddly moving. He suspects they mean he was moving oddly.

A champion of non-locality, he is never where he is supposed to be. He began *The Quantum Universe* while playing the part of Gulliver, continued it in a field dressed as a Middle Age monk on location for a BBC costume drama; and completed it when working on Interconnectedness at Princeton University. His Unpredictability, unnervingly evident at all times, makes it horribly difficult to know what he is going to tackle next. Or even whether he will finish it; a matter which also raises problems of Uncertainty and Continuity.

He firmly believes that everyone has a book inside them. That this one was in there was just his bad luck.